INCREDIBLY DISGUSTING ENVIRONMENTS™

POLLUTION AND YOUR LUNGS

Daniel E. Harmon

rosen publishing's
rosen
central®

New York

Published in 2013 by The Rosen Publishing Group, Inc.
29 East 21st Street, New York, NY 10010

First Edition

Library of Congress Cataloging-in-Publication Data

Harmon, Daniel E.
Pollution and your lungs/Daniel E. Harmon.—1st ed.
     p. cm.—(Incredibly disgusting environments)
Includes bibliographical references and index.
ISBN 978-1-4488-8410-0 (library binding)—ISBN 978-1-4488-8425-4 (pbk.)—
ISBN 978-1-4488-8426-1 (6-pack)
1. Lungs—Diseases—Prevention. 2. Lungs—Dust diseases—Prevention. 3. Air—Pollution—
Health aspects. 4. Health promotion. I. Title.
RC756.H37 2013
616.2'405—dc23

                                                                    2012026582

Manufactured in the United States of America

CPSIA Compliance Information: Batch #W13YA: For further information, contact Rosen Publishing, New York, New York, at 1-800-237-9932.

CONTENTS

INTRODUCTION

To live is to breathe—literally. Life requires oxygen. For good health, humans and other life-forms need clean air to breathe. Bad air can cause big problems.

The air today is not as clean and healthy as it once was. In many parts of the world, people can not only smell the air, they can see and even feel it. It is so thick and massive in certain areas due to pollution that astronauts in space can see it and have taken pictures of it. To people on the ground, it sometimes obscures the sun. Pollution is gross—and very dangerous.

Air quality has worsened especially in the last 150 years. The Industrial Revolution, which began in the mid-1800s, started a period of wonderful progress for society. It brought about automation, new inventions, and great advances in lifestyles, but it also increased pollution. The Industrial Revolution was based on factories powered by coal heating. At the time, most people did not realize the burning of coal—a fossil fuel—sends polluting smoke into Earth's atmosphere. Smoke, to them, was just the stuff that rose from chimneys and vanished into the sky. No one gave it serious thought.

Scientists know now that smoke does not simply disappear. It lingers in the atmosphere until it is eliminated by natural processes. Trees and other plants can absorb pollutants that linger. The bad news is that plant habitats, particularly Earth's rain forests, are being reduced by companies that want their timber and land area. At the same time, more new pollutants are entering the atmosphere.

Major modern-day pollutants are carbon emissions from factories, automobiles, and heating systems. Cigarette smoke is a leading form of pollution indoors. Some types of air pollution occur naturally. They include volcanic eruptions and other geological activity, as well as wildfires.

A smog alert greets motorists on a busy interstate highway in Atlanta, Georgia. Such cautions are common around large cities, especially in summer.

Weather can impact the air. Smog is a combination of smoke and fog. It also describes the type of pollution that results when heavy volumes of auto emissions in large cities react with sunlight. This produces ground-level ozone. It creates a visible canopy of particles that constantly stays in the area, giving the sky above a brownish color. At times, smog has become so thick that cities basically have been forced to shut down. Schools, public offices, and businesses have closed. Hospitals have been overwhelmed by thousands of patients with respiratory problems—some of them critical and even fatal.

Smog and other types of pollution are disgusting to see. What they do to the human body is even more disgusting. Bad air causes lung and respiratory illnesses and disorders. It can also worsen heart and lung problems that people already have.

Coughing, breathing difficulties, and inflamed nasal passages are the most obvious symptoms of irritation from air pollutants. Over time, physical damage may become life threatening. Pollution is linked to cancer and other diseases. Because the air people breathe exchanges with the bloodstream, pollution can also affect the circulation.

Everyone needs to explore the causes of pollution and its effects on the lungs. Young people should also look at things that are being done to address the problem and ways in which they can contribute to solutions.

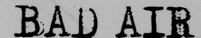

1 BAD AIR

A human breathes almost 3,000 gallons (11,356 liters) of air in a day, on average. A person's brain function begins to shut down if deprived of oxygen-containing air for more than three or four minutes. With continued lack of oxygen, other organs in the body can begin to shut down, too.

Pollution in the air does not ordinarily cause death in minutes, but it can be just as deadly in a more subtle way. Unfortunately, people may not realize the dangers in the air they are breathing. Harmful effects due to air pollution may not become evident in the human body for a few years or even decades, although the damage has been done and can be irreversible.

What Is Air Pollution?

Smog, a combination of human-made pollutants and weather conditions, is common around large industrial centers. Some

people have come to think of smog and pollution as the same thing, but smog is just one form of air pollution.

Leading ingredients of air pollution include the following:

• Ground-level ozone. The U.S. Environmental Protection Agency (EPA) explains that ozone is "good up high, bad nearby." Ozone in the upper atmosphere is necessary to protect Earth from harmful solar rays.

Towering, belching smokestacks were once signs of industrial progress. Today they are causes for environmental concern as they release inestimable quantities of pollutants into the atmosphere.

Ground-level (tropospheric) or "bad" ozone is the basic type of smog. Both "high" and "low" ozone have the same chemical makeup.

- Particulate matter. Examples of visible particles are soot and ash from fires, diesel smoke, tiny metal fragments, and chemicals like nitrates, sulfates, and carbon compounds. Ozone and particles are the two forms of pollution most dangerous to humans, according to the American Lung Association.

- Carbon monoxide. This slightly lighter-than-air gas is produced by the incomplete burning of certain kinds of fuel. After entering the atmosphere, it combines with oxygen, resulting in carbon dioxide. Carbon monoxide is difficult for humans to detect because it has no smell, taste, or color.

- Nitrogen dioxide. This poisonous gas is a combination of nitrogen and oxygen. It is highly reactive, which means it can be stimulated easily by other substances and processes. Most nitrogen oxides cannot be seen or smelled, but nitrogen dioxide combined with air particles produces the amber atmospheric tint visible around highly polluted locales.

- Sulfur dioxide. Another toxic gas, sulfur dioxide results from burning sulfur or compounds that contain it. Sulfur is found, for example, in petroleum and coal. Like nitrogen dioxide, sulfur dioxide is highly reactive.

- Lead. Lead is a chemical element. In metallic form, it is used in the manufacture of many products—building material, ammunition, electrical batteries, etc. Too much lead in the human body comprises a dangerous neurotoxin.
- Ammonia. A compound of hydrogen and nitrogen, ammonia has a strong, pungent smell. While useful in pharmaceuticals, fertilizer, cleaners, and other products, certain types and levels of exposure to it can endanger health.

It's a little scary to realize how many dangerous chemicals or harmful gases can be in polluted air. Most of the time, you can't even see or smell them. Fortunately, the EPA has measures in place to detect these agents and warn the community if dangerous levels occur.

Causes of Air Pollution

The production processes used in large factories create both solid and atmospheric waste. The sight of smoke belching from towering smokestacks is a particularly disgusting environmental image. However, numerous other substances pollute the air.

Ground-level ozone is created when the burning of fossil fuels produces gases that interact with sunlight. Summer months naturally are the worst for generating ground-level ozone. Primary sources of emissions that result in ozone are electric utilities, auto exhaust, fuel vapor, chemical solvents, and aerosols.

POLLUTION STATISTICS

It is impossible to know how many people die of pollution-related causes, but scientists consider it a leading killer. David Pimentel, a Cornell University ecologist, in 2007 estimated that sixty-two million people die each year as a result of exposure to environmental hazards, notably pollution of the water, air, and soil.

Dirty water, which can carry all sorts of life-threatening infections, is the main worry. Air pollution is another leading concern. The World Health Organization has attributed as many as three million deaths annually to respiratory diseases, many of them caused or provoked by air pollution.

Pimentel calculated that Americans have more than a hundred foreign (not naturally occurring) chemicals in their bodies. Many of them are poisons.

Particle pollution is created mainly by wood- and coal-burning stoves and industrial furnaces. It is also produced by diesel fuel used in cars, trucks, and heavy construction and agricultural equipment.

The cause of hazardous automobile emissions is simple: the combustion of gasoline. When the engine is running in a gas-powered vehicle, waste is produced and emitted through the exhaust pipe. Exhaust contains carbon monoxide and other hydrocarbons as well as lead.

Nitrogen dioxide is caused mainly by highway and street traffic and the combustion of various fossil fuels. Indoors, some of the sources of nitrogen dioxide are kerosene heaters, gas stoves, and faulty combustion appliances. Tobacco smoke and welding also produce nitrogen dioxide.

Sulfur dioxide mostly comes from combustion processes at power plants. Other industrial operations produce smaller amounts. When trains, ships, and heavy equipment burn fuels with a high sulfur content, this type of emission occurs. It is also produced during ore-mining processes.

A leading source of lead in the atmosphere, in the United States, is the processing of ores and metals. Although lead content is no longer allowed in automobile gasoline in this country, it is still used in aviation fuel. Lead is also released into the air when waste is burned.

Ammonia that pollutes the air is produced mainly in farming. It is a key ingredient of fertilizers used to stimulate crop growth. Ammonia is also contained in a variety of cleaning fluids.

There are many more sources of atmospheric pollutants. They include military and other explosives, nuclear elements escaping after disasters at power plants, slash-and-burn agricultural practices, emissions from waste disposal sites, emissions from household heating systems, and careless garbage incineration.

Pollutants rise from gas, oil, and chemical leaks. Methane gas comes from garbage landfills and from animal waste. Methane is most commonly associated with the greenhouse effect in the upper atmosphere and global warming, but it is also linked to ground-level ozone.

Cigarette smoking is a common form of air pollution both indoors and out.

Not all pollution is created by humans. Natural events and processes have contaminated the air since ancient times. They include volcanoes, natural gas emissions from rock fissures, smoke from wildfires started by lightning strikes, dust storms, and pollen.

Besides endangering humans, air pollution affects plants and animals.

Indoor Pollution

Many people think of the "environment" as the great outdoors. "Environmental pollution," to them, is outdoor pollution. Pollution indoors, though, can actually be more deadly.

Germs thrive in houses and office buildings that have poor filtering and circulation. Germs in the air aren't the only problem, though. Other sources of indoor air pollution include smoke from stoves and fireplaces, secondhand tobacco smoke, asbestos, and pesticides, as well as some of the major pollutants discussed earlier (lead, carbon monoxide, etc.).

A category of ailments called "sick building syndrome" can happen when people are exposed to poor air quality and circulation. A person is literally sickened by breathing in the germs and/or poisons that have accumulated in the place where he or she works or lives.

Radon gas, which occurs naturally within the earth, seeps upward and sometimes collects in home basements. Formaldehyde gas, although not considered a major contributor to indoor pollution, is also suspected of being a carcinogen (cancer-causing agent). Formaldehyde is contained in glues, common household products, and some types of pressed wood used in home interiors.

In its basic forms, indoor pollution is a problem especially in underdeveloped countries. Millions of people in those regions live in badly ventilated houses and burn wood, coal, dead plants, and even animal dung for heating. Smoke fills the houses.

The World Health Organization has estimated that 1.6 million people die each year of indoor air pollution—twice as many as the fatalities from outdoor pollution. The reason is that indoor pollutants can be inhaled in much higher concentrations.

Medics transport a teacher from a middle school where numerous cases of apparent "sick building syndrome" resulted from a mysterious source inside the buildings.

MYTHS & FACTS

MYTH: People are safe from air pollution if they don't spend much time outside.

FACT: According to the American Lung Association, "Indoor air can be even more polluted than the air outdoors." Homes and other buildings that are poorly ventilated are especially unhealthy. So are older buildings that were built with materials such as asbestos, once common in construction but now known to be hazardous.

MYTH: Young people don't need to worry as much about air quality as grown-ups.

FACT: Air pollution has a faster and more dangerous impact on smaller individuals. Children are more greatly affected than adults by ground-level ozone for several reasons. Their lungs are not fully developed, and they are more prone to asthma. Also, children generally spend more time outdoors. Health officials say air pollution is a major contributing factor in premature deaths.

MYTH: Government regulations during the past thirty years are solving the pollution problem effectively.

FACT: It's true that certain forms of air and other pollution have been reduced greatly as a result of strict environmental requirements. Generally, though, much still needs to be done. Some cities and regions are worse off today than ever before.

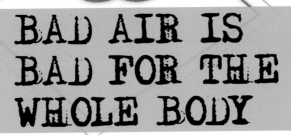

2 BAD AIR IS BAD FOR THE WHOLE BODY

Pictures of cars, trucks, and factories

polluting the air are disgusting enough. The alarming health statistics that result are beyond disgusting.

Worldwide, air pollution is one of the ten leading causes of premature deaths. The World Health Organization estimates one and a half billion people around the world are breathing seriously bad air. Many scientists and medical researchers fear it will worsen in coming years.

In March 2012, the Organisation for Economic Cooperation and Development (OECD) published a report titled "OECD Environmental Outlook to 2050: The Consequences of Inaction." During the next forty years, the group forecasts, the world population will increase dramatically. So will living standards. Energy demand will increase by 80 percent—primarily in developing countries—and the OECD predicts that 85 percent of the energy will continue to be produced from fossil fuels.

That means air quality will deteriorate further. Pollution-related health problems will multiply. The OECD believes that 3.6 million people worldwide will be dying of "particulate

air pollutants [ground-level ozone] leading to respiratory failure"
by the year 2050. Big city air pollution may become the greatest
environment-related contributor to premature deaths—more
dangerous than filthy water and poor sanitation, which long have
been the main threats.

Bad Air Versus Human Lungs

The effects of air pollution on the lungs are so gross they actually can
be seen. To illustrate, doctors have removed and exhibited damaged

Particulate matter from auto exhausts in Hong Kong forces pedestrians to cover
their mouths and noses on a day of high pollution.

and undamaged lungs from the corpses of organ donors. The polluted specimens, badly discolored, can be identified at first glance.

Coloration only hints at what goes on inside the body of an individual plagued by polluted air. Pollution impairs the constant action of the lungs and affects the entire respiratory system. Dr. Campion E. Quinn, author of *100 Questions & Answers About Chronic Obstructive Pulmonary Disease*, points out, "The lungs and airways are highly sensitive to inhaled irritants. Prolonged exposure to inhaled irritants can lead to inflamed and narrowed air passages that make it difficult to inhale and exhale."

Air pollution aggravates existing lung disorders and diseases. These include asthma, emphysema, tuberculosis, pneumonia, bronchitis, and even the common cold. Pollution can cause dangerous complications. At the same time, air pollution can cause new disorders to develop in healthy individuals.

Lung tissue removed from a deceased person contains black areas caused by soot and other particulates as well as passive smoking. The individual was a nonsmoker.

Atmospheric pollutants, such as ground-level ozone or harmful gases and particles, enter the human lungs through normal breathing. From there, via the bloodstream, they spread throughout the body. The lungs are at the greatest risk, but other organs are at risk as well.

Ground-level ozone is what people living in polluted metropolitan areas breathe constantly. Even small ozone intakes can damage the body.

KILLER SMOG

In most cases, pollution kills slowly. Medical victims of bad air are weakened over a period of time and die of related diseases. Sometimes, though, fatal pollution occurs suddenly—when people cringing in a burning house, fighting a forest fire, or caught in the vicinity of a volcanic eruption are overcome by smoke and vapors. In other circumstances, pollution develops rapidly but not so dramatically.

Merriam-Webster's Collegiate Dictionary defines "fog" as a "vapor condensed to fine particles of water suspended in the lower atmosphere that differs from cloud only in being near the ground." When elements of human-produced air pollution blend with fog, the resulting smog can be deadly.

More than 4,000 Londoners died of fog/smog-related causes over a period of two weeks in December 1952. In the darkest hours, it literally was impossible to see a hand held in front of the face. History repeats itself. In December 1991, 160 people died in London, apparently as a result of a four-day spell of smoggy, breezeless weather. This tragic phenomenon has not been limited to London with its famous "pea soup" fogs. Similar events have occurred in different parts of the world.

Children especially are at risk because their respiratory, circulatory, and cardiovascular systems have not developed completely. The greatest lung growth occurs between the ages of ten and eighteen years. In a study sponsored by the National Institute of Environmental Health Sciences, researchers found that the pollutants in fossil fuel emissions can hurt the normal development of children's lungs. The result may be a lifetime of subnormal breathing—less than 80 percent the capacity of healthy lungs.

Skin irritations and inflammation of the nasal passages may be the only discomforts, at first. Over time, exposure to air pollutants can inflame the lungs, affect the rhythms of the heart, and lead to life-threatening health problems. For instance, excess nitrogen dioxide in the air immediately irritates the eyes, nose, and throat. In

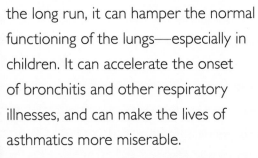

the long run, it can hamper the normal functioning of the lungs—especially in children. It can accelerate the onset of bronchitis and other respiratory illnesses, and can make the lives of asthmatics more miserable.

Airborne particle pollutants can trigger asthma attacks and even cause heart attacks or strokes. In addition to breathing problems, lead pollutants

While cigarette smoking is obviously highly dangerous for smokers, secondhand smoke threatens nonsmokers who inhale it. That is why smoking is prohibited in many public places.

may contribute to heart problems, kidney diseases, fertility problems, weakened immune systems, and irritation of the nervous system.

Less widespread pollutants introduce their own baggage of ill effects. For example, the breathing of radon gas—evident in some home and commercial building basements—can contribute to lung disease. Formaldehyde gas, although not such a serious or long-term peril in most circumstances, presents serious health problems to some people if it occurs at high levels in small, enclosed quarters. Airborne formaldehyde can irritate the respiratory tract. Watery eyes may evolve into coughing, tightness of the chest, nausea, and skin rashes. Ammonia in the air can irritate and damage the skin and, if inhaled, the internal organs.

Tobacco smoke is a common airborne pollutant that damages the lungs and other systems in the human body—and it affects not just the smoker but also people nearby who breathe the smoke. Secondhand smoke inhaled over a period of time can cause serious illnesses in nonsmokers.

What's Happening to the Whole Body?

The damage that air pollution causes human tissues is more than disgusting. It's dismaying and hideous. And it's far-reaching. It extends from the nasal cavities into the lungs and into internal organs and vital systems.

RESPIRATORY SYSTEM

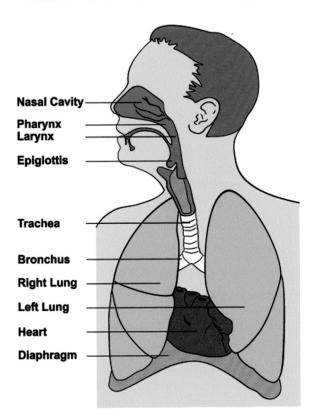

Nasal Cavity

Pharynx
Larynx

Epiglottis

Trachea

Bronchus

Right Lung

Left Lung

Heart

Diaphragm

Damage begins with the nose—or even sooner. Some individuals are affected by certain pollutants immediately, on contact. Even before they inhale a poison, it irritates their skin.

The air humans inhale, whether clean or polluted, first passes through the nasal cavities behind the nose and mouth. Through the pharynx, larynx, and trachea, the air enters the lungs, and from there, the inner realms of the body. Every stop along the way (beginning with the exterior skin) is

This diagram shows the different parts of the upper respiratory system through which air is inhaled and exhaled. Each area can be damaged by pollution.

susceptible to contamination. If the air breathed in has been polluted, it might affect any component of the upper respiratory tract.
Eventually, it will also affect internal organs that receive oxygen
processed and supplied by the lungs.

In sum, polluted air affects the entire respiratory system and
beyond. It touches not only air passages and lungs. It can damage
many parts of the human body.

Avoidable Pollution Tragedies

It's interesting that many health crises could easily be avoided.
The U.S. Consumer Product Safety Commission reports that
about 170 American citizens die each year from carbon monoxide poisoning. Carbon monoxide commonly comes from car
emissions. People have died from lack of oxygen when cars
are left running too long in closed garages because of the high
production of carbon monoxide without adequate ventilation.
Other sources of carbon monoxide poisoning come from home
furnaces, stovetops, fireplaces, electric-powered room heaters,
water heaters, portable generators, and more. Poisoning from
portable generators, for example, caused forty-seven confirmed
deaths in the aftermaths of violent weather events in 2005 (the
year of Hurricane Katrina).

3

SYMPTOMS, DISEASES, AND DISORDERS

Immediate physical reactions to air pollution are irritations: burning eyes, dry or inflamed nasal passages and throat, and coughing. They can become more worrisome: shortness of breath, fatigue, chest pains, and blurred vision. Symptoms of common illnesses such as the cold are more severe for people living in heavily polluted cities.

The early symptoms can go away if a person gets out of the polluted surroundings. Over time, though, the damage may become permanent if the person continues to breathe the polluted air. Air pollution can worsen serious medical conditions that exist. It also can bring about new illnesses and disorders.

Lung and Respiratory Diseases and Disorders

Because breathing is the main way air pollution enters the human body, the areas of the body most at risk are the lungs

COPD: A GROWING GLOBAL TRAGEDY

COPD (chronic obstructive pulmonary disease) is a combination of conditions that limit a person's airflow. Emphysema and chronic bronchitis are the two major diseases that are classified as COPD. As with asthma, shortness of breath is one symptom, but while medication can overcome asthma attacks, it cannot eliminate all the discomforts of COPD. The disorder progressively results in the loss of lung performance.

Some fourteen million Americans are believed to suffer from COPD. Some medical analysts predict that worldwide, by 2020 it may become the third most common cause of death. Patients with COPD can live for many years, but the disease greatly diminishes their quality of life. They lack physical energy. Many require oxygen therapy in addition to medication and become work-disabled while still in their prime.

Smoking is the primary cause of COPD in the United States. In developing countries, smoke from household cooking and heating fires is a major pollutant.

and respiratory system. Some illnesses and disorders are relatively mild and treatable. Others are often fatal.

- Asthma. This disorder of the airways brings about chronic coughing and wheezing, often with labored breathing and tightness of the chest. Asthma attacks

may come on without warning because of the sufferer's hypersensitive air passages.

Research has shown that ground-level ozone not only intensifies asthma but may also actually cause it in otherwise healthy individuals. In a five-year study, scientists at the University of Southern California found that children living in an ozone-dense area who played outdoor sports were several times more likely to develop asthma than were the nonathletes.

Asthmatics must take their inhalers wherever they go to counter an onset of the ailment. Asthma is common among young people, whose respiratory systems are still developing.

- Beryllium disease. Beryllium is a metal used in manufacturing a variety of products, from dental bridges to golf clubs to computer parts. Dust from beryllium mines and processing plants pollutes the air, soil, and water. If breathed, it can cause mild respiratory problems. Chronic beryllium disease scars the lungs and can be fatal. If a person is exposed to the dust over a long period, the risk of lung cancer increases.

- Black lung, also called "cotton worker's lung," "farmer's lung," and "flock worker's lung," is a breathing condition that commonly affects coal miners. Over a period of time, coal dust lodges in the lungs, causing the lung tissue and smaller airways to harden. In advanced stages, coughing is frequent and breathing can be difficult and painful. Black lung disease has led to countless premature deaths.

- Silicosis, a form of black lung disease, can result from long overexposure to inhaled silica dust and causes similar lung functioning problems.

- Byssinosis, or "cotton worker's lung," usually occurs in textile workers. It comes from breathing the dust of cotton, hemp, flax, and other vegetable fibers. Coughing, wheezing and tightness of the chest are the main symptoms. They generally diminish when the person is away from the workplace for a few days.

- Farmer's lung is a lung inflammation caused by breathing dust, mold, and fungus of the kind usually found on farms—straw, hay, and grain dust. Fever, chills, coughing, and breathing problems can occur within hours after inhaling a large amount of dust. If exposed over an extended period, the person may develop a chronic lung disease.

- Asbestosis. People who have been exposed to very

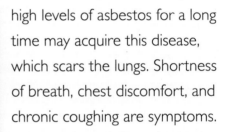

high levels of asbestos for a long time may acquire this disease, which scars the lungs. Shortness of breath, chest discomfort, and chronic coughing are symptoms.

A particular problem with asbestosis is that early detection is practically impossible. It may be as long as twenty years after the exposure period before symptoms occur—and they resemble the symptoms of numerous other respiratory ailments.

- Flock worker's lung, despite the term, is not caused by working with chickens or other birds. Rather, it generally occurs among people who work with synthetic fabric

Blackened, thickened scar tissue is obvious on the lungs of this eighty-six-year-old victim of asbestosis. Problems likely began to develop many years before the disease was diagnosed.

fibers called flock. Symptoms vary, but they usually include a persistent dry cough and labored breathing sometimes accompanied by chest pains.

- Bronchitis. The bronchi are the larger airways in the lungs. Bronchitis inflames the lining of the bronchi, called mucous membranes. Irritative bronchitis can be caused by breathing ammonia, sulfur dioxide, and other harmful fumes. Air pollution is especially blamed for acute bronchitis, which may last several months. Chronic bronchitis, especially that associated with tobacco smoking, may last for years and in many cases is classified as a form of COPD.

- Emphysema. Emphysema is another form of COPD that results from damage to the lungs' tiny air sacs. Air becomes trapped inside the sacs, making it hard for a person to breathe out the old air fully or to breathe in new air. This means the person cannot obtain sufficient oxygen, and higher levels of poisonous carbon dioxide, normally expelled by the lungs, remain in the lungs and blood.

- Pneumonia. Pneumonia is an infection of the lung tissues that can cause breathing difficulty and can even be fatal. Most pneumonia is due to bacteria and viruses, but smoking and other forms of indoor air pollution can make it worse. People in regions where particulate matter is a problem and people with existing lung conditions are more likely to need hospitalization for pneumonia than people elsewhere. Pneumonia is the leading cause of death among the elderly. Additionally, the World Health Organization reports that almost half the deaths from acute lower respiratory infections among infants and young children "are due to particulate matter inhaled from indoor air pollution from household solid fuels."

- Respiratory allergies. These allergies affect the eyes, skin, and air passages. Symptoms vary from mild to severe. They include watery or irritated eyes, itchiness in the nose and throat, coughing and wheezing, and breathing difficulty. Indoor pollution is blamed for the increase in occurrences of these types of allergies in recent years.

- Sleep apnea. Heavy snoring and brief interruptions of breathing while asleep are common signs of sleep apnea. (Not everyone who snores has apnea.) Apnea is a pause in regular breathing, which might occur a dozen or more times in an hour. The person with this disorder is usually unaware of it, but the condition can be serious. The brain automatically triggers the lungs to resume breathing eventually, but sleep apnea has been known to cause sudden death. Smoking increases the risk of the disorder.

- Lung cancer. Cigarette smoking is highly associated with various cancers of the lung tissue. Medical scientists note that the risk of lung cancer increases for smokers who have also had long-term exposure to asbestos. Mesothelioma, a cancer of the linings around the lungs, is caused mainly by exposure to asbestos. Nonsmokers (including those not exposed to secondhand smoke) can also develop lung cancer, although no one currently knows exactly why.

A medical scan shows that this woman's left lung is missing. She is a survivor of cancer that doctors believe was caused by asbestos exposure when she was a child.

Diseases and Disorders in Other Parts of the Body

The list of nonrespiratory illnesses caused or worsened by air pollution is almost endless. Pollution is rarely the direct cause of these ailments, but it can aggravate the symptoms. As a result, people living in areas of heavy air pollution have to rely more on medications for these diseases and require more doctor visits and hospitalizations.

Heart disease is the most dangerous condition that can be affected by air pollution. Some researchers, in fact, suspect air pollution poses greater threats to the heart than to the lungs. They note that more heart attacks, strokes, and hospitalizations for other heart ailments occur on days when air pollution levels are up. What happens is that pollutants—most notably particulate matter—enter the bloodstream after being inhaled. Among other effects, they worsen clogged arteries, which eventually can lead to strokes and heart attacks.

Although research is inconclusive, some medical scientists suspect air pollution may also be a trigger for headaches, including tension headaches and migraines. Changes in air pressure are known to affect chronic headache sufferers. Pollution can also adversely affect pregnant women. For example, babies may be born dangerously underweight.

10 GREAT QUESTIONS
TO ASK AN ENVIRONMENTALIST

1. How is it that some people live all their lives in densely polluted areas and aren't affected by pollution? They're active into their eighties or nineties and finally die of unrelated causes.

2. In a heavily polluted metropolitan area like Los Angeles, how long does it take, on the average, for pollution to permanently affect the respiratory system?

3. Are their possible links between air pollution and diabetes? Obesity? Mental or emotional illnesses?

4. Is the disgusting damage to the lungs caused by cigarette smoke reversible?

5. I live in a notorious smog zone. Should I wear a breathing mask when I go outside on especially bad days in summer?

6. Can COPD be cured?

7. Do air-conditioning systems effectively remove pollutants that enter a home or office building from outside?

8. Why do some industrial and heavily populated areas have worse smog than others the same size?

9. What is the most harmful air pollutant?

10. If more trees are planted in areas of heavy air pollution, can they reduce the problem significantly?

4

IMPROVING THE AIR

Better air is the key to healthier lungs.
Finding ways to improve air quality should be an endeavor of everyone who breathes it. Many young people believe there is little they can do, but they should know that they can take action to become part of the solution.

What Students Can Do

Students and their parents can discuss with school officials ways to improve indoor air quality (IAQ). Some schools already have formed IAQ committees. In those that haven't, concerned students can press for their creation. They can also spearhead the formation of clean-air student clubs, with the mission of improving air quality both inside school buildings and around the city. When appropriate, they can choose pollution-related themes for research papers and science projects. In this way, they can better educate themselves about the problems of air contamination and, in doing so, educate others.

Families with asthmatic children should talk to the school's nurses and teachers to make sure exposure to indoor asthma triggers is minimized.

At home, young people can join parents in cleaning up the air indoors. Some of the steps families can take include these:

- Establish a smoke-free home. Do not permit visitors to smoke indoors. If a smoker is a member of the family, encourage the person to kick the habit; at the very least, insist that smoking be done only outdoors.
- Think twice before lighting scented candles to get rid of stinky odors. It's better to locate and remove the source of the smell than to cover it up.
- Research and shop for the least toxic "green" household cleaning products.
- Don't let a car engine idle in the garage.
- If the kitchen is equipped with a gas stove and no venting system is installed, have a fan installed with an exhaust vent leading outside.

Environmentally conscious homemakers are using natural cleaning substances in their everyday tasks. Reducing the presence of manufactured cleaning products is one way to combat pollution.

- Use dehumidifiers to keep the indoor humidity level below 50 percent. Make sure the dehumidifiers work properly—otherwise, they can add to the pollution.
- Correct plumbing leaks. Wet places foster mold and other sources of pollution that enter the air.
- When painting or using hobby or craft supplies that contain poisonous chemicals, make sure the room is well ventilated.
- Be sure the house is guarded by carbon monoxide detectors as well as smoke detectors. Remind parents to have the home checked periodically for radon gas levels.
- When parents undertake renovations and home improvement projects involving compressed wood, point out that they should use woods that emit the slightest amounts of air pollutants—preferably, alternatives to products containing formaldehyde. Solid wood or metal is better.
- In cold weather, have central heating systems and fireplaces inspected by professionals. Make sure chimney flues are open when wood is burning in fireplaces. Use vented, not unvented, space heaters. Make sure wood stoves comply with the EPA's emission guidelines; see that their doors and other openings close snugly.

Parents have the ultimate authority in establishing family policies and making home maintenance and improvement decisions.

Children can and should express their opinions, though, especially when the family's health is at stake.

Government Initiatives

The United States passed the Clean Air Act in 1963. Amendments, especially those added in 1970, have strengthened it. The purpose of the law, in short, is to protect citizens from air pollution. Some environmentalists question how effective it has been in achieving that; they urge tighter emission regulations. Statistics indicate, though, that remarkable reductions in certain types of air pollutants are obvious. Other countries have enacted similar regulations.

At lower government levels, cities, counties, and states have attempted to control emissions in highly polluted areas. Many cities have imposed antismoking ordinances. Millions of restaurants, malls, and other public facilities across the country forbid cigarette smoking.

Government agencies in the United States and other countries have developed air quality indices (AQIs). The U.S. Environmental Protection Agency provides daily AQI updates on the Internet concerning air quality in different parts of the nation. Color-coded charts quickly show whether the air quality in a particular metropolitan area or county is good, moderate, unhealthy for sensitive groups, generally unhealthy, very unhealthy, or hazardous.

Environmental groups and private companies, meanwhile, are promoting innovations in natural energy development (sun, wind, and water power). If these sources of energy can be made effective and deployed on a wide scale, carbon emissions from fossil fuels can be reduced.

Efforts to Reduce Auto Emissions

For many years, environmentalists and government agencies have urged Americans to drive less. The Environmental Protection Agency says driving cars is the single most harmful thing Americans do to worsen air pollution. Gasoline-powered transportation is a necessary part of life for most Americans at present, but there are ways to reduce auto pollution. Regular tune-ups keep the engine running as efficiently as possible. Proper tire inflation improves gas mileage. When it is time to trade, consumers should consider hybrids and other alternative autos that produce less toxic emissions.

The best way to reduce, or at least slow, increasing pollution from automobiles is to drive only when necessary and to carpool and use mass transportation whenever possible. (A happy upshot of Internet advances is that more and more workers are telecommuting—working at home. They can perform and submit their work to employers hundreds or thousands of miles away, or on other continents. In the past, they drove many miles each day to work and back. Now, several days may pass before they need to drive at all.)

AIR-FRIENDLY AUTOMOBILES

Public interest in reducing the amount of gasoline used in automobiles did not stem from environmental concerns. It was all about economics. In the winter of 1973–1974, political events that centered in the oil-producing Middle East resulted in a calamitous shortage of gas at the pumps in the United States. Gas had to be rationed, and gas prices soared. Americans were forced to find ways to limit the amount of fuel they bought. Automakers came up with car models that got better mileage, while inventors experimented with alternatives to gasoline-powered autos.

Since then, progress has been slow but significant. Three major alternatives now exist to the old-fashioned gas-guzzling American automobile: 1) new cars designed to get better mileage, 2) hybrids, which are cars that can run some of the time on battery power, recharged by regular gasoline power, and 3) all-electric cars.

Cars that can go more than 40 miles (64.4 kilometers) on a gallon of gas have been available for many years, but automakers seem more focused on customers' demands for interior space, stylishness, comfort, and extras. Thus, 30 miles (48.3 km) to a gallon is now considered very good mileage.

Hybrids are capturing increased public interest. A great initial concern was that hybrid manufacturers offered only limited warranties on the engine batteries—and those batteries are very expensive to replace.

So far, total-electric cars have proved suitable in limited situations, where cars are driven only short distances. (And recharging the battery, some point out, requires fossil fuel–based electrical current.)

More Must Be Done

The good news is that air quality in many ways has improved as a result of the Clean Air Act in the United States and similar regulations in other countries. For example, the EPA reports that pollution by carbon monoxide in the United States decreased by more than 50 percent during the first decade of this century. The level of lead pollution in the air decreased by almost 90 percent between 1980 and 2010.

Still, air pollution persists and, in many areas, is worsening. The key to making air safer to breathe is education. People of all ages should stay informed about pollution facts and statistics and do their part to hold pollutants in check.

A rental car company in New York provides this charging station for renters of its electric cars. Electric autos are viable alternative vehicles in short-distance driving situations.

GLOSSARY

aerosol Liquid or solid particles suspended in mist form.

carbon dioxide Chemical compound formed of carbon monoxide and oxygen (also known as CO2).

carbon monoxide Gas formed by the partial burning of fossil fuels (also known as CO).

carcinogen Cancer-causing substance.

cardiovascular Relating to the heart and blood vessels.

combustion Process of burning.

fossil fuel Oil, coal, or other fuel derived from the remains of ancient plants and animals.

hydrocarbon Compound of hydrogen and carbon found in coal, petroleum, and other fossil fuels.

Industrial Revolution Period of major, basic changes in manufacturing, mining, transportation, farming, and technology processes; the American Industrial Revolution occurred in the eighteenth and nineteenth centuries.

mucous Relating to mucus, a moist, slippery respiratory secretion.

neurotoxin Poison that affects the nervous system.

ozone Bluish oxygen form with a pungent smell, necessary for blocking solar rays in the upper atmosphere but harmful in the lower atmosphere.

pesticide Chemical used in farming and gardening to control insects.

FOR MORE INFORMATION

American Lung Association

1301 Pennsylvania Avenue NW, Suite 800

Washington, DC 20004

(202) 785-3355

Web site: http://www.lung.org

This major nonprofit organization is dedicated to improving lung health through educational programs, research, and advocacy projects.

The Canadian Environmental Network (RCEN)

39 McArthur Avenue, Level 1-1

Ottawa, ON KIL 8L7

Canada

(613) 728-9810

Web site: http://rcen.ca/home

RCEN promotes "networking among environmental organizations and others who share its mandate to protect the Earth and promote ecologically sound ways of life."

National Institutes of Health (NIH)

9000 Rockville Pike

Bethesda, MD 20892

(301) 496-4000

Web site: http://www.nih.gov

A government organization, NIH seeks "fundamental knowledge about the nature and behavior of living systems and the application of that knowledge

to enhance health, lengthen life, and reduce the burdens of illness and disability."

U.S. Environmental Protection Agency (EPA)

Ariel Rios Building

1200 Pennsylvania Avenue NW

Washington, DC 20460

(202) 272-0167

Web site: http://www.epa.gov

This government agency has overall responsibility for monitoring the environment and enforcing regulations.

World Health Organization (WHO)

Avenue Appia 20

1211 Geneva 27

Switzerland

+ 41 22 791 21 11

Web site: http://www.who.int/en

A United Nations authority, the WHO directs and coordinates health issues.

Web Sites

Due to the changing nature of Internet links, Rosen Publishing has developed an online list of Web sites related to the subject of this book. This site is updated regularly. Please use this link to access the list:

http://www.rosenlinks.com/IDE/Lungs

FOR FURTHER READING

Bailey, Jacqui. *What Happens When You Breathe?* (How Your Body Works). New York, NY: Rosen Publishing, 2009.

Casper, Julie Kerr. *Fossil Fuels and Pollution: The Future of Air Quality* (Global Warming). New York, NY: Facts On File, 2010.

Curley, Robert, ed. *New Thinking About Pollution* (21st Century Science). New York, NY: Britannica Educational Publishing, 2010.

Desonie, Dana. *Atmosphere: Air Pollution and Its Effects* (Our Fragile Planet). New York, NY: Chelsea House Publishers, 2007.

Feinstein, Stephen. *Solving the Air Pollution Problem: What You Can Do* (Green Issues in Focus). Berkeley Heights, NJ: Enslow Publishers, 2010.

Ferreiro, Carmen. *Lung Cancer* (Deadly Diseases & Epidemics). New York, NY: Chelsea House Publishers, 2007.

Greenhaven Press. *Lung Cancer* (Perspectives on Diseases & Disorders). Farmington Hills, MI: Greenhaven Press, 2010.

Parker, Russ. *Pollution Crisis* (Planet in Crisis). New York, NY: Rosen Publishing, 2009.

Rapp, Valerie. *Protecting Earth's Air Quality* (Saving Our Living Earth). Minneapolis, MN: Lerner Publishing, 2008.

Sechrist, Darren. *Air Pollution* (Saving Our World). Tarrytown, NY: Marshall Cavendish Children's Books, 2008.

Sheen, Barbara. *Lung Cancer* (Diseases & Disorders). Farmington Hills, MI: Lucent, 2010.

BIBLIOGRAPHY

Grosvenor, Carrie. "Air Pollution Facts." LoveToKnow/Green Living. Retrieved April 2012 (http://greenliving.lovetoknow .com/Air_Pollution_Facts).

Harvard Medical School. "Health Effects of Air Pollution." Harvard Health Publications. August 2005. Retrieved April 2012 (http://www.health.harvard.edu/press_releases /health_effects_air_pollution).

Harvey, Fiona. "Air Pollution 'Will Become Bigger Global Killer Than Dirty Water.'" *The Guardian*, March 15, 2012. Retrieved April 2012 (http://www.guardian.co.uk /environment/2012/mar/15/air-pollution-biggest-killer-water).

Hecht, Barbara K., and Frederick Hecht. "Air Pollution Stunts Kids' Lungs." MedicineNet. Retrieved April 2012 (http://www.medicinenet.com/script/main/art .asp?articlekey=38983).

Jenkins, McKay. *What's Gotten Into Us? Staying Healthy in a Toxic World*. New York, NY: Random House, 2011.

Judd, Sandra J., ed. *Respiratory Disorders Sourcebook*. 2nd ed. (Health Reference Series). Detroit, MI: Omnigraphics, 2008.

National Institute of Environmental Health Sciences. "Air Pollution and Respiratory Disease." Retrieved April 2012 (http://www.niehs.nih.gov/health/impacts/respiratory).

Quinn, Campion E. *100 Questions & Answers About Chronic Obstructive Pulmonary Disease.* Sudbury, MA: Jones and Bartlett Publishers, Inc., 2006.

Secor, Melinda L. "Types of Air Pollution." LoveToKnow/Green Living. Retrieved April 2012 (http://greenliving.lovetoknow.com /Types_of_Air_Pollution).

Skye, Jared. "Air Pollution Statistics." LoveToKnow/Green Living. Retrieved April 2012 (http://greenliving.lovetoknow.com/Air _Pollution_Statistics).

Thompson, Andrea. "Pollution May Cause 40 Percent of Global Deaths." LiveScience.com, September 10, 2007. Retrieved April 2012 (http://www.livescience.com/1853-pollution-40-percent -global-deaths.html).

Weil Weekly Bulletin. "Pollution and Migraine." January 13, 2011. Retrieved April 2012 (http://www.drweil.com/drw/u /WBL02167/Pollution-and-Migraine.html).

World Health Organization. "Indoor Air Pollution and Health." September 2011. Retrieved April 2012 (http://www.who.int /mediacentre/factsheets/fs292/en).

INDEX

About the Author

Daniel E. Harmon has written more than eighty books, including works on various health, medical, and environmental topics and profiles of the U.S. Environmental Protection Agency and other government branches. Works recently published include *Al Gore and Global Warming* and *Jobs in Environmental Cleanup and Emergency Hazmat Response*. He is the author of thousands of magazine and newspaper articles and editor of a national computer technology newsletter. Harmon lives in Spartanburg, South Carolina.

Photo Credits